# The Occult Anatomy of Man

## TO WHICH IS ADDED A TREATISE ON OCCULT MASONRY

### BY

### Manly P. Hall

---
SIXTH EDITION
---

Martino Publishing
Mansfield Centre, CT
2013

Martino Publishing
P.O. Box 373,
Mansfield Centre, CT 06250 USA


ISBN 978-1-61427-434-6


© 2013 Martino Publishing

All rights reserved. No new contribution to this publication may
be reproduced, stored in a retrieval system, or transmitted, in any form or
by any means, electronic, mechanical, photocopying, recording, or otherwise,
without the prior permission of the Publisher.


Cover design by T. Matarazzo




Printed in the United States of America On 100% Acid-Free Paper

# The Occult Anatomy of Man

TO WHICH IS ADDED A TREATISE
ON OCCULT MASONRY

BY

MANLY P. HALL

---
SIXTH EDITION
---

PHILOSOPHICAL RESEARCH SOCIETY
3341 GRIFFITH PARK BLVD., LOS ANGELES, CAL.
1937

*Copyrighted Dec. 10, 1929*
*By Manly P. Hall*
*For permission to copy or translate,
address the author.*

# THE OCCULT ANATOMY OF MAN

## THE HUMAN BODY IN SYMBOLISM

### Part I

In Scripture we are told that God made man in His own image. It is so stated not only in the Christian Bible but also in the holy writings of nearly all enlightened people. The Jewish patriarchs taught that the human body was the Microcosm, or little Cosmos, made in the image of the Macrocosm, or the great Cosmos. This analogy between the finite and the Infinite is said to be one of the keys by the aid of which the secrets of Holy Writ are unlocked. There is no doubt that the Old Testament is a physiological and anatomical textbook to those capable of reading it from a scientific viewpoint. The functions of the human body, the attributes of the human mind, and the qualities of the human soul have been personified by the wise men of the ancient world, and a great drama has been built around their relationships to themselves and to each other. To the great Egyptian demigod, Hermes, the human race owes its concept of the law of analogy. The great Hermetic axiom was, *That which is above is like unto that which is below, and that which is below is like unto that which is above.* The religions of the ancient world were all based upon Nature-worship, which in a degenerated form has survived to our own day as phallicism. The worship of the parts and functions of the human body began in the later Lemurian period. During the Atlantean epoch this religion gave place to sun-worship, but the new faith incorporated into its doctrines many of the rituals and symbols of the previous belief. The building of temples in the form of the human body is a custom common to all peoples. The Tabernacle of the Jews, the great Egyptian Temple of Karnak, the religious structures of the Hawaiian priests, and the Christian churches laid out in the form of the Cross, are all examples of this practice. If the human body were laid out with the arms spread in the shape of any of these buildings, it would be found that the high altar would occupy the same relative position in the building that the brain occupies in the human body.

All the priests of the ancient world were anatomists. They recognized that all the functions of Nature were reproduced in miniature in the human body. They therefore used man as the textbook, teaching their disciples that to understand man was to understand the universe. These wise men believed that every star in the heavens, every element in the earth, and every function in Nature was represented by a corresponding center, pole or activity within the human body.

This correlation between Nature without and the nature of man within, which was concealed from the multitudes, formed the secret teachings of the ancient priestcrafts. Religion in Atlantis and Egypt was taken much more seriously than it is today. It was the very life of these peoples. The priests had complete control over the millions of ignorant men and women who had been taught since childhood that these robed and bearded patriarchs were direct messengers from God; and it was believed that any disobedience to the commands of the priests would bring down upon the offender's head the wrath of the Almighty. The temple depended for its maintenance upon its secret wisdom, which gave its priests control over certain powers of Nature and made them vastly superior in wisdom and understanding to the laity whom they controlled.

These wise ones realized that there was a great deal more involved in religion than the chanting of mantrams and the singing of hymns; they realized that the path of salvation could be walked successfully only by those who had practical, scientific knowledge of the occult function of their own bodies. The anatomical symbolism which they evolved in order to perpetuate this understanding has come down to modern Christianity, but the keys to it have apparently been lost. It is a tragic situation for religionists that they are surrounded by hundreds of symbols which they cannot understand; but it is still sadder that they have even forgotten that these symbols ever had any meaning other than the foolish interpretations which they themselves have concocted.

The idea prevalent in the minds of Christians that their faith is the one and only truly inspired doctrine, and that it came parentless into the world, is unreasonable in the extreme. A study of comparative religions proves beyond doubt that Christianity has begged, borrowed or stolen its philosophies and concepts from the religions and philosophies of the ancient and mediaeval pagan worlds. Among the religious symbols and allegories which belonged to the world long before the coming of Christianity are a few to which we would like to call your attention. The following Christian symbols and concepts are of pagan origin:

The Christian cross comes from Egypt and India; the triple mitre from the faith of the

Mithraics; the shepherd's crook from the Hermetic Mysteries and Greece; the Immaculate Conception from India; the Transfiguration from Persia; and the Trinity from the Brahmins. The Virgin Mary, as the mother of God, is found in a dozen different faiths. There are over twenty crucified world Saviors. The church steeple is an adaptation of Egyptian obelisks and pyramids, while the Christian devil is the Egyptian Typhon with certain modifications. The deeper one goes into the problem, the more he realizes that there is nothing new under the sun. A truly sincere study of the Christian faith proves beyond all doubt that it is the evolutionary outgrowth of primitive doctrines. There is an evolution in religion as well as in physical form. If we accept and incorporate into our doctrines the religious symbolism of nearly forty peoples, it behooves us to understand (at least in part) the meaning of the myths and allegories which we borrow, lest we be more ignorant than those from whom we secured them.

This brochure is devoted to the problem of explaining the relationship existing between the symbolism of the ancient priests and the occult functions of the human body. We must first understand that all sacred writings are supposed to be sealed with seven seals. In other words, it requires seven complete interpretations to understand fully the meaning of these ancient philosophic revelations which we have liked to call Holy Writ. Scripture is not intended to be historical. Those who understand its literal meaning understand the least of its meaning. It is a well-known fact that for dramatic reasons in his plays Shakespeare brought characters together who had actually lived hundreds of years apart; but Shakespeare was not writing history—he was penning drama. The same is true of the Bible. Scripture leaves historians hopelessly involved in self-contradictory chronological tables, where the majority of historians will remain until the judgment day. Scripture furnishes excellent subject matter for debates and also ground for hair splitting over the meaning of terms and probable locations of unknown cities. Most of the Bible towns now pointed out by guides were named hundreds of years after the birth of Christ by pilgrims who suspected them of having occupied sites somewhere near those mentioned in the Bible. All this may prove convincing to some, but to the thinker it is conclusive evidence that history is the least important part of Scripture.

When the Empress Helena visited Jerusalem in 326 A. D., the mother of Constantine the Great discovered that not only all trace of Christianity had already been lost, but that a temple dedicated to the Goddess Venus stood on the hill now accepted as Mount Calvary. Less than four hundred years after the death of Christ there was apparently no one in the Holy Land who had ever heard of Him! This does not necessarily imply that He never lived, but it certainly does indicate that the halo of miracles and supernatural atmosphere with which modern Christianity envelops Him are largely mythological. Like all other religions, the Christian faith accumulated a weird collection of fantastic legends which are its own worst enemies; for they have taken the simple moralist of Nazareth—the man who loved His fellow creatures—and built around Him a superstructure of idolatry which loves no one and serves only itself.

As Buddha in India merely reformed the Brahmin concepts of his day, so Jesus re-shaped the faith of Israel and gave to His disciples and the world a doctrine based upon that which had gone before but remodeled to meet the needs of the people who surrounded Him and the problems which confronted the Jewish nation. The Essenes who educated Jesus were of Egyptian or Hindu origin, and His faith incorporated the best of that which had gone before. The records preserved of Him are largely allegorical, and the simple man is plunged by them into a great sea of supernaturalism. This was not entirely without purpose, however, for as Shakespeare took license with history in order to present essential truths, so it seems the historians of Jesus used the character of the man as the groundwork of a great drama. He becomes the hero of a seven-sealed story, and those Christians who have studied symbolism can gain from that story the key of the true Christian Mysteries. They will then realize that Scripture is perpetual history; that it pertains to no nation or people but is the story of all nations and peoples. It is a wonderful thing, for example, to study the life of Christ in the light of astronomy, for He becomes the sun, and His disciples the twelve signs of the zodiac. Among the constellations we find the scenes of His ministry, and in the precession of the equinoxes the story of His birth, growth, maturity, and death for men. Again, the tortured chemicals in the retort further reveal to us the life of the Master, for with the key to chemistry the Scriptures become another book. In this particular work, however, we can only concern ourselves with the relationship of these allegories to the human body.

The life of Christ, as found in the Gospels, we discover, has been conventionalized until it agrees perfectly with the lives of dozens of world Saviors, for all of them are also astronomical and physiological myths. All of these myths come to us out of the most remote antiquity, where the primitive races used the human body as the symbolic unit and the gods and demons were personified out of the organs and functions of the body. Among certain Qabbalistic writers the Holy Land is mapped out on the human body and the various cities are shown as centers of consciousness in man. There is a wonderful study here for those who will investigate deeply and sincerely the ancient Mysteries. We cannot hope to cover all the ground, but if you can gain from this booklet a key to the situation, we hope you will pursue the line of thought until you have made it all-inclusive and opened at least one seal of the Book of Divine Revelation.

# THE THREE WORLDS
## Part II

According to the Mystery Schools, the human body is divided into three major parts; and in analogy with this the universe without was said to be composed of three worlds: namely, heaven, earth, and hell. Heaven was the superior world and for some unknown reason was supposed to have been above, although Ingersoll proves conclusively that, owing to the rotation of the earth, *up* and *down* are always changing places. Nearly all religions teach that God dwells in the heavens. Their members are taught to believe that God is above them, so they raise their hands in prayer and lift their eyes to the heavens when they implore or petition Him. Among some nations He is supposed to dwell on the tops of mountains, which are the highest places of the world. Wherever He is and whatever He is, His place of domicile is above, overshadowing the world below.

Between heaven above and hell beneath is the place which the Scandinavians call Midgard, or the middle garden. This is called the earth. It was suspended in space and formed the dwelling place of men and other living creatures. It was connected to the heavens by a rainbow bridge down which the gods descended. Its volcanic craters and fissures were said to connect it with hell, the land of darkness and oblivion. Here, "twixt heaven and earth dominion wielding," as Goethe said, exists Nature. The green grass, the flowing rivers, the mighty ocean exist only in the middle world, which is a sort of neutral ground where the hosts of good and evil fight their eternal battle of Armageddon.

Below, in darkness and flames, torment and suffering, is the world of Hel, which we have interpreted as hell. It is the great beneath; for as surely as we think of heaven as up we think of hell as down, while this middle place (earth) seems to be the dividing line between them. In hell are the forces of evil, the tearing, rending, destroying powers, which are always bringing sorrow to the earth and which struggle untiringly to overthrow the throne of the gods in heaven.

The entire system is an anatomical myth, for the heaven world of the ancients—the doomed temple on the top of the mountains—was the skull with its divine contents. This is the home of the gods in man. It is termed *up* because it occupies the northern end of the human spine. The temple of the gods who rule the earth is said to be at the North Pole, which also, by the way, is the home of Santa Claus, because the North Pole represents the positive end of the spinal column of the planetary lord. Santa Claus coming down the chimney, with his sprig of evergreen (the Christmas tree), at the season of the year when Nature is dead has a fine Masonic interpretation for those who wish to study it. The same is true of the manna that descended to feed the Children of Israel in the wilderness, for this manna is a substance which comes down the spinal cord from the brain. The Hindus symbolized the spine as the stem of the sacred lotus; therefore the skull and contents are symbolized by the flower. The spinal column is Jacob's Ladder, connecting heaven and earth, while its thirty-three segments are the degrees of Masonry and the number of years of the life of Christ. Up these segments the candidate ascends in consciousness to reach the temple of initiation located on the top of the mountains. It is in this domed room with a hole in the floor (*foramen magnum*) that the great mystery initiations are given. The Himalaya mountains rise above the earth, representing the shoulders and upper half of the body. They are the highest mountains of the world. Somewhere upon their summits stands the temple, resting (like the heavens of the Greeks) upon the shoulders of Atlas. It is interesting to note that the Atlas is the upper vertebra of the human spine upon which the condyles of the skull rest. In the brain there are a number of caves (ventricles and folds), and in them (according to Eastern legends) live the wise men—the Yogis and hermits. The caves of the Yogi are said to be located at the head of the Ganges river. Every religion has its sacred river. To the Christians it is the Jordan; to the Egyptians is was the Nile; while to the Hindus it is the Ganges. The sacred river is, of course, the spinal canal, which has its source among the peaks of the mountains. The holy men in their retreats represent the spiritual eye sense centers in the human brain. These holy men are the seven sleepers of the Koran, who must remain in the darkness of their cave until the spirit fire vitalizes them.

The brain is, of course, the upper room referred to in the Gospels where Jesus met with His disciples, and it is said that the disciples themselves represent the twelve convolutions of the brain. It is these twelve convolutions which later send their messages by means of the nerves into the body below to convert the Gentiles, or preach the gospel in the middle earth. These twelve convolutions gather around the central opening in the brain (the third ventricle), which is the Holy of Holies

—the Mercy Seat—where between the spreading wings of the angels Jehovah talks with the high priest and where both day and night the Shekinah's glory hovers. From this point also the spirit finally ascends from Golgotha, the place in the skull. It is a clairvoyant fact that the spirit not only leaves but also enters the body through the crown of the head—probably giving rise to the story of Santa Claus and his chimney.

The Trinity in man lives in the three great chambers of the human body, from which they radiate their power throughout the three worlds. These centers are the brain, the heart, and the reproductive system. These are the three main chambers of the pyramid and also the rooms in which are given the Entered Apprentice, Fellowcraft, and Master Mason's degrees of Blue Lodge Masonry. In these three chambers dwell the Father, the Son, and the Holy Ghost, who are symbolized by the three-lettered word, AUM. The transmutation, regeneration, and unfoldment of these three great centers results in the sounding of the *Lost Word,* which is the great secret of the Masonic Order From the spinal nerves come impulses and life forces which make this possible. Therefore the Mason is told to consider carefully his substitute word, which means "the marrow of the bone."

In the cerebellum, or posterior brain—which has charge of the motive system of the human body and is the only brain developed in the animal—is to be found a little tree-like growth which has long been symbolized as a sprig of acacia and as such is referred to in the Masonic allegory.

The two lobes of the cerebrum were called by the ancients Cain and Abel, and have much to do with the legend of the curse of Cain, which is literally the curse of unbalance. For the murder of the spirit of equilibrium, Cain is sent forth a wanderer upon the face of the earth. I have in my possession a very remarkable skull which originally rested on the shoulders of a homicide. It is of high organic quality, but bears the curse of Cain. This individual had a grudge which he nursed very carefully. Nursed grudges sometimes become very dangerous things. This person swore that when he met a certain man he would cut his heart out and throw it in his face. A number of years passed, his hatred grew, and at last meeting his enemy, he attacked him and fulfilled his threat. He was hanged for the crime, but the skull bearing the testimony to the brain, reveals a very interesting fact. The right half of the brain is under the control of Mercury—the planet of intelligence—and as a result of the crossing of the brain nerves at the base of the skull it rules the left side of the body. The left half of the brain, under the control of Mars—the spirit of anger and impulse—rules the right side of the body and likewise the strong right arm. As the result of his hatred and the rulership of Mars which grew out of that hatred, the left rear side of the brain is fully twice the size of the right side. The individual allowed Mars to control his nature. The impetuosity of Mars ruled him, and he paid with his life for the mark of Cain. Science knows there is a very narrow line between genius and insanity; for any dominating vice or virtue man must pay with unbalance. Unbalance always distorts the viewpoint, and distorted viewpoints are unfailingly productive of misery.

In the skull is the switchboard which controls the activities of the body. Every function of man below the neck is controlled by a center of consciousness in the brain. Proof of this is the fact that injury to certain centers of the brain results in paralysis of various parts of the body. Medical science now knows that the spinal cord is an elongation of the brain, and some authorities even claim the cord to be capable of intelligence throughout its entire length. This cord is the flaming sword which is supposed to have stood at the gates of the Garden of Eden. The Garden of Eden is in the skull, within which is a tree bearing twelve manner of fruit.

The brain is filled with vaulted chambers and passageways which have their correspondence in the spans and arches of the temples, while the third ventricle is undoubtedly the King's chamber of the Great Pyramid. The spinal cord is the serpent of the ancients. In Central and South America the Saviour God is called Quetzalcoatl. His name means a *feathered serpent,* and this has always been his symbol. This is the brazen serpent raised by Moses in the wilderness. The nine rattles on the tail of the serpent are called the number of man and they represent the sacral and coccygeal bones, within whose centers the secret of human evolution is contained.

Every organ of the physical body is reproduced in the brain, where it can be traced by the law of analogy. There are two embryonic human forms, one male and the other female, twisted together in the brain. These are the Yin and the Yang of China, the black and white dragons biting at each other. One of these figures has as its organ of expression the pineal gland, and the other the pituitary body. These two ductless glands are well worth consideration, for they are very important factors in the unfolding of human consciousness. While appearing to have no functions, they have not atrophied; and, as Nature preserves no unnecessary organs, they must have a very important part to play. It is known that these glands are larger and more active in higher grades of mentality than in those of lower quality, and in certain congenital idiots they are very small. These two little glands are called the head and the tail of the dragon of wisdom. They are the copper and zinc poles of an electric circuit which has the entire body as a battery.

The pituitary body (which rests in the sella turcica of the sphenoid bone directly behind and just a little below the bridge of the nose and connected to the third ventricle by a tiny tube called the infundibulum) is the feminine pole, or negative center, which has charge of the expressions of physical energy. Its activity also regulates to a large degree the size and weight of the body. It is also a thermometer revealing disorder in any other of the chain of ductless glands. Endocrinology (the study of the ductless glands and their secretions) is still in its embryonic stage, but some day it will be revealed as the most important of all medical sciences. The pituitary body is known under the following symbols by the ancient world: The alchemical retort, the mouth of the dragon, the Virgin Mary, the Holy Grail, the lunar crescent, the laver of purification, one of the Cherubim of the Ark, the Isis of Egypt, the Radha of India, and the fish's mouth. It may well be called the hope of glory of the physical man. At the opposite end of the third ventricle and a little higher is the pineal gland, which looks not unlike a pine cone (from which it secured its name).

E. A. Wallis Budge, keeper of the Egyptian antiquities in the British Museum, mentions in one of his works the Egyptian custom of tying pine cones on the tops of their heads. He states that in the papyrus rolls these cones are fastened to the tops of the heads of the dead when taken into the presence of Osiris, Lord of the Underworld. Undoubtedly this symbol referred to the pineal gland. It was also the custom of certain African tribes to fasten pieces of fat to the tops of their heads, and allow it to melt in the sun and run down over them, as part of their religious observances. It is interesting that the American Indian should wear his feather—which was originally symbolical of his Christ—in the same place where the Christian monk shaves his head. The Hindus teach that the pineal gland is the third eye, called *the Eye of Dangma*. It is called by the Buddhists *the all-seeing eye,* and is spoken of in Christianity as *the eye single.*

We are told that ages ago the pineal gland

was an organ of sense orientation by which man cognized the spiritual world, but that with the coming of the material senses and the two objective eyes, it ceased to be used and during the time of the Lemurian race retreated to its present position in the brain. It is said that children, recapitulating in childhood their previous periods of evolution, have limited use of the third eye up to their seventh year, at which time the skull bones grow together. This accounts for the semi-clairvoyant condition of children, who are far more sensitive than adults along psychic lines. The pineal gland is supposed to secrete an oil, which is called *resin*, as the life of the pine tree. This word is supposed to be involved in the origin of the *Rosicrucians*, who were working with the secretions of the pineal gland and seeking to open the eye single; for it is said in Scripture, "If thine eye be single, thy body shall be filled with light."

The pineal gland is the tail of the dragon and has a tiny finger-like protuberance at one end. This gland is called *Joseph*, for it is the father of the God-man. The finger-like protuberance is called *the staff of God*, sometimes *the holy spear*. Its shape is like the evaporating vessels of the alchemist. It is a spiritual organ which is later destined to be what it once was: namely, a connecting link between the human and the divine. The vibrating finger on the end of this gland is *the rod of Jesse* and *the sceptre of the high priest.* Certain exercises as given in the Eastern and the Western Mystery Schools cause this little finger to vibrate, resulting in a buzzing, droning sound in the brain. This is sometimes very distressing, especially when the individual who experiences the phenomena, in all too many cases, knows nothing about the experiences through which he is passing.

In the middle of the brain and surrounded by the convolutions is the third ventricle—a vaulted chamber of initiation. Around it sit three kings, three great centers of life and power—the pituitary body, the pineal gland, and the optic thalamus. In this chamber also is a small gritty seed, which is undoubtedly connected with the king's coffer in the Great Pyramid. The third ventricle is supposed to be the seat of the soul, and the aura radiating from the heads of saints and sages is said to represent the golden glow pouring from this third ventricle.

Between the eyes and just above the root of the nose is a spreading in the frontal bone of the skull, which is called the frontal sinus. The slight bulge caused by the spreading of this bone is known to phrenology as the seat of individuality. It is here that the jewels are placed on the foreheads of the Buddhas, and it is also from this point that the serpent rose from the crown of the ancient Egyptians. Several of the Mystery Schools teach that this is the seat of Jehovah in the human body. While His function is through the generative system, His center of consciousness as a part of the spirit of man was supposed to be located in a sea of blue ether called *the veil of Isis*, in the center of the frontal sinus. When clairvoyantly studying the body of man, that little point always shows up as a black dot and cannot be analyzed.

The Palatine Hill of the ancients, upon which were built the temples of Jupiter and Juno, also has its place in the human body. The palate bone is a sort of hill-shaped structure, and right above it are the orbital cavities containing the two eyes, which are the Jupiter and Juno of the ancient world.

The cross, of course, represents the human body. The upper limb of it is the head of man rising above the horizontal line of his outstretched arms. As already stated, the great churches and cathedrals of the world have been built in the form of a cross, and contain (where the head should be) the altar, upon which are burning lighted candles. These candles are symbolic of spiritual sense centers in the brain, while the custom of placing a rose window over the altar suggests the soft place in the top of the skull. The skull—the upper room—is the sanctum sanctorum of the Masonic Temple, and to it only the pure can aspire.

The winged bone, which medical science knows as the sphenoid, is the Egyptian scarab, carrying in his claws the pituitary body and also bearing aloft the gleaming spark of immortality located in the frontal sinus.

We are told in ancient mythologies that the gods came down from heaven and walked with men, instructing them in the arts and sciences. In a similar way the godlike powers in man descend from the heaven world of his brain to carry on the work of constructing and reconstructing natural substances. We are told that in the ultimate of evolution man's body will slowly be dissolved back again into the brain (which was its origin) until nothing remains but seven globular centers radiating seven perfect sense perceptions, which are the Spirits before the Throne and the Saviors which he is bringing into the world to redeem it through the seven periods of his growth.

Man is an inverted plant, gaining his nourishment from the sun as the plant does from the earth. So, as the life of the plant ascends its stem to nourish its leaves and branches, the life of man (rooted in the brain) descends to produce the same result. This life descending is symbolized as the World Saviors who come down into the world to die for men. Later these lives are returned again to the brain, where they glorify man before all the worlds of creation. So much for the story of the brain. Now let us consider the next of man's marvelous parts: namely, the spinal column.

# THE SPINAL COLUMN
## Part III

Connecting the two worlds (heaven above and the sphere of darkness below) is the spinal column—a chain of thirty-three segments protecting within them the spinal cord. This ladder of bones plays a very important part in the religious symbolism of the ancients. It is often referred to as a winding road or stairway. Sometimes it is called *the serpent;* at other times, *the wand* or *sceptre.*

The Hindus teach that there are three distinct canals or tubes in the spinal system. They call them *Ida, Pingala,* and *Sushumna.* These tubes connect the lower generative centers of the body with the brain. The Greeks symbolized them by the Caduceus, or winged staff of Hermes. This consisted of a long rod (the central Sushumna), which ended in a knob or ball (the pons of the medulla oblongata). On each side of this knob are wings arched over, used to represent the two lobes of the cerebrum. Up this staff and twisted around it wind two serpents, one black and the other white. These represent the Ida and Pingala.

The ancient Hindus have a legend concerning the Goddess Kundalini, in which it is said that she descended by means of a ladder or cord from heaven to a little island floating in the great sea. Connecting this with embryology, it is evident that the ladder or cord represents the umbilical cord, while the little island is the solar plexus. When the ladder is cut away from heaven, the goddess flees in terror to a cave (sacral plexus), where she hides far from the sight of men. Like Amaterasu, the Japanese Goddess of the Shining Face, she must be lured from her cave, for while she is there and refuses to come forth the world is in darkness. *Kundalini* is a Sanskrit word meaning "a serpentine or twisting gas or force." This force (so the Eastern sages claim can be drawn up through the central spinal canal (the Sushumna). When this essence strikes the brain, it opens the centers of spiritual consciousness and inner perception, bringing with it spiritual illumination. The system of culture whereby this is made possible is the most secret teaching of the Eastern saints, for they realize that this spiral, twisting force is not only illuminating but (like the serpent which is its symbol) also a deadly poison. Smatterings of Eastern occultism are coming all the time into the Western world; but we are sorry to say they are bringing with them endless suffering and sorrow, for these great truths in the hands of individuals incapable of rightly understanding or applying them destroy intelligence and reason. Along the spine are a number of nerve ganglia and plexus. All of these have their place in religious symbolism. For example, we are told that the early Jews called the sacral plexus and the sacrococcygeal ganglion *the cities of Sodom and Gomorrah.* There is a small plexus in the region of the kidneys called *the sagittarial plexus,* which the ancients knew as *the city of Tarsus,* where St. Paul fought with the beasts. Higher occultism teaches that the lotus blossoms (nerve centers on the spine) are reflections or negative poles bearing witness of seven great positive centers of consciousness located in the brain. These seven function through the centers on the spine in approximately the same way that the Seven Spirits before the Throne function through the planetary bodies. The disciple is warned not to work with the centers upon the spine, but to labor instead with their true rulers—the centers in the brain. The wandering of the Children of Israel in the wilderness, the pilgrimage of the Mohammedan to Mecca, the endless pilgrimages of Hindu holy men who spend their lives wandering from one shrine to another, represent the pilgrimage of the spirit fire (*Kundalini*) through the nerve centers along the spine. By certain specifications, force is turned into these centers one after the other until, when seen clairvoyantly, they are great, flower-like areas from which light rays stream like petals. Each of these lotus blossoms has a different number of petals, according to the number of nerves which branch out from it.

It is said that the Logos, when the time came to create the material universe, entered a state of deep meditation, centralizing His thought power upon the seven flower-like centers of the seven worlds. Gradually His life force descended from the brain (which was the great superior world) and, striking these flowers one after the other, gave birth to the lower worlds. When at last His spirit fire struck the lowest center, the physical world was created and His fire was at the base of the spine. When the world returns to Him again and He once more becomes supreme in consciousness, it will be because He withdraws the life from these seven centers, beginning with the lowest, and returns it again to the brain. Thus the path of evolution for all living things is to raise this fire, whose descent made their manifestation in these lower worlds possible and whose raising brings them into harmony once more with the superior worlds. This myth of the life force that came down and took upon itself worlds is found among all the civilized nations of the earth. This is the Hiram Abiff who built the Masonic temple (the bodies) and was then slain by the three vehicles which he formed. This is likewise the Christ, slain for the sins of the world.

Because of the fact that this spinal fire is a twisting, serpentine force, the snake has been used in all parts of the world to represent the World Saviors. The Uræus worn by the Egyptian priests upon their foreheads (like the brazen serpent raised by Moses in the wilderness) was symbolic of *Kundalini*—the sacred cobra—who, when she was raised in the wilderness, saved all who gazed upon her (Moses and the brazen serpent).

As the brain is the center of the divine world, so the solar plexus is the center of the human world, for, representing semiconsciousness, it links the unconsciousness below with the consciousness above. Man is not only capable of thinking with his brain; he is capable of a certain phase of thought through the nerve centers of the solar plexus. It will probably be wise at this point to describe the difference between a medium and a clairvoyant. To the average person there is no difference, but to the mystic these two phases of spiritual sight are separated by the entire span of human evolution.

A clairvoyant is one who has raised the spinal serpent into the brain and by his growth earned the right of perceiving the invisible worlds with the aid of the third eye, or pineal gland. This organ of consciousness, which millions of years ago connected man with the invisible worlds, closed during the Lemurian period when the objective senses began to develop. The occultists, however, by this process of development hinted at before, may reopen this eye and by means of it explore the invisible worlds. Clairvoyants are not born; they are made. Mediums are not made; they are born. The clairvoyant can become such only after years, sometimes lives, of self-preparation; on the other hand, the medium, by sitting in a darkened room or other similar practices, may secure results in a few days.

The medium uses the solar plexus as a mirror, and upon its sensitive nerves are reflected pictures recorded in the invisible ethers. Through the spleen (which is the gateway to the etheric body) the medium permits decarnate intelligences to come into his spiritual constitution, the result being voices and other spirit manifestations. Automatic writing is gained by permitting the etheric arm of an outside intelligence to control temporarily the physical arm of the medium. This is not possible until the medium removes his own etheric double from the arm, for two things cannot occupy the same place at the same time. The result of separating the life forces from the physical arm periodically is very disastrous, often resulting in paralysis. Mediumship is unnatural to man, while clairvoyance is the natural result of growth and the unfolding of the spiritual nature. There are a hundred mediums to one clairvoyant, for the clairvoyant can become such only through self-mastery and the exertion of tremendous power; while the weaker, the more sickly and the more nervous an individual is, the better medium he makes. The clairvoyant is unfolding his mind by filling it with useful knowledge, while the first instruction given the would-be mediums is, "Make your mind a blank."

The reason why mediumship through the solar plexus is a retrogression, may be summed up as follows: The Group Spirits who control the destinies of the animal kingdoms govern their charges through pictures thrown against the solar plexus, for the animal has no self-conscious mind. As a result, instead of thinking with its own brain, it thinks with the brain of the Group Spirit to whom it is attached by invisible magnetic cords. These cords convey their impressions and photograph them upon the sympathetic nervous system. Having no will of its own, the animal is incapable of combating these urges and consequently obeys them implicitly. Man governs himself through the cerebro-spinal nervous system. Man, however, has developed individuality and the sympathetic system no longer rules him. Opening himself to impulses through the solar plexus area, the medium is therefore thwarting his own growth by preventing the cerebro-spinal nervous system from controlling his destiny.

Man has always liked to lean on external things. He hates to face each problem and solve it with the brain God gave him. He consequently leans on the invisible worlds, asking them to help him to accomplish the labors which he should do by his own efforts.

Thousands of people must carry the Karmic responsibilities of the medium, for many follow this calling because hundreds of people want to talk to deceased relatives or get inside tips on the stock exchange. Those who by their patronage encourage things of which they do not themselves approve are personally responsible for the injury which their selfishness has thus permitted other persons to inflict upon themselves. The difference, therefore, between mediumship and clairvoyance is about half the length of the spinal column. It is the difference between negative and positive; it is the difference between the midnight of the seance room and the noonday of the temple.

All of the organs within the body of man have their religious significance. The heart, with its chambers, is itself a temple standing upon the mountain of the diaphragm. The spleen, with its little umbrella-shaped body, collects the sun's rays and has charge of the etheric body. It is this etheric body coiled within the spleen that injects the white blood corpuscles into the circulatory system.

We know that the human body has been the inspiration for nearly all mechanical contrivances. Hinges are copied from the human body; likewise, the ball and socket joint. We are told that the first plumbing was reproduced from the arterial and venous circulatory systems. Hundreds of machines and implements which work around us have been inspired by the subtle workings of our own vehicles, for the human body is the most marvelously constructed machine which the human mind can study.

The close relationship between the generative system below and the brain above (for the brain is a positive generating system) is, of course, due to the spine connecting them. At a certain time, a number of little doors, which now separate the brain from the generative system, are opened and the Sushumna becomes an open tunnel so that every impulse is carried immediately to both ends of the body. It is for this reason that the candidate assumes the vows of celibacy, for the close connection existing in the advanced disciple between the brain and the reproductive system necessitates an absolute conservation of all life energies. The tonsils are directly connected with the generative system; in fact, are part of its positive pole in the brain. The present deplorable custom of vaccinating children and cutting out their tonsils the moment they appear in the world will some time result in a distinct degeneration of the race. Most tonsils are infected as the result of the child eating too many sweets during the first few years of life. The moral is, Don't cut out the tonsils; cut out the sweets. Most parents are responsible for the sickness of their children. Through either ignorance or indulgence they allow the infantile consciousness, as yet uncontrolled by its higher vehicles, to destroy itself before life fairly begins. When children are sick during the early years of life, the physician will usually find the cause of the ailment in the parent; and the father or mother—and not the child—should be dosed with pills. If the stomach is kept in proper condition, the tonsils will give very little trouble. The absolute economy exhibited by Nature in the building of all its structures should be sufficient proof that the Lord was not wasting his time when He made tonsils and appendices. He apparently had a reason for making them, but these poor unoffending organs have become a gold mine to medical scientists, who remove them at the slightest provocation. We are told that the vertical position assumed by the human body, which thereby forces the contents of the intestinal tract to travel part of the time uphill, is the contributing cause of the appendix, which is missing in creatures of horizontal carriage. Every organ not only has its visible purpose but it also has an invisible spiritual purpose, and the individual is to be envied who manages to go through life preserving intact as many as possible of his original anatomical parts and members.

While on the subject of the debt of science to the human body, we might add that the decimal system is the result of the primitive man counting on his fingers, whereby ten became the unit of enumeration. The ancient cubit also was the distance between the elbow and the end of the second finger, or approximately eighteen inches. So it goes back through the study of things until we find that nearly everything with which man has surrounded himself in an adaption from the body with which God surrounded his spirit.

Man is gradually gaining control not only over the organs of his body but also over their functions. Science states that certain organs function automatically or mechanically, but occultism realizes that there is nothing mechanical about the functions of the human body. Let us take as an example a workman throwing a piece of iron among the wheels and levers of a smoothly working machine. There is a grinding crash and the machine stops. If, on the other hand, you throw a monkey wrench figuratively into the human body, it would immediately begin the process of throwing it back at you. It would surround the foreign element with a coating and would try to absorb it. If this were impossible, it would try to eject it through some channel appointed for that purpose. If this means failed, it would in many cases accustom itself to the presence of the obstacle and keep right on working anyway. This shows unmistakably that the organic parts of man possess some inherent form of intelligence; therefore they are not machines, for no mechanical device is capable of intelligence.

Paracelsus, the great Swiss physician, who after many years in the far East returned to Switzerland to teach medicine, first gave to the European world its concept of the Nature Spirits. Paracelsus taught that the functions of Nature were under the control of little creatures, invisible to the normal senses but who, working through the kingdoms of life, minerals, plants, animals, and parts of the human body, kept all of these evolving in an intelligent way. Under the control of the great Celestial Hierarchy of Scorpio, which has charge of all body-building in Nature, these elementals are the invisible intelligences governing the human body and its functions.

As the result of man's ever evolving consciousness, he is acquiring more complete control over the functions of his various organs. There are two kinds of muscles—voluntary and involuntary—the difference being that the voluntary muscles which are controlled by the conscious mind of the individual have fibres running two ways and crossing each other, while those which are involuntary are without cross fibres. The heart used to be considered an involuntary muscle, but it is now beginning to show cross fibres, thus foreshadowing the day when man will consciously and intelligently regulate the beating of his own heart. The same will be true with respect to all other organs that survive the periodic changes taking place in the constitution of man. The Eastern holy man can successfully live without his heart beating; he can stop it or start it at will. By lifting the tongue so that it closes the air passages into the lungs, he can remain in suspended animation for months without dying. Many Eastern *chelas* do this while being given spiritual initiations out of the physical body. There are cases on record where these holy men have been buried in the ground. Weeks later when the body was dug up it was found to be dried like leather. Water was poured on it, and after a certain lapse of time the man who had not taken a breath in weeks got up and walked off. This is the result of extraordinary control which the mind is capable of gaining over the functions of the body.

Occultism teaches that there is an entire universe within the human body—that it has its worlds, its planes, and its gods and goddesses. Millions of minute cells are its inhabitants. These are grouped together into kingdoms, nations, and races. There are the bone cells and the nerve cells and millions of these tiny creatures grouped together become one thing composed of many parts. The Supreme Ruler and God of this great world is the consciousness in man which says "I am." This consciousness picks up its universe and moves to another town. Every time it walks up and down the street it takes a hundred million solar systems with it but, being so infinitesimal, man cannot realize that they are actually worlds.

In like manner, we are individual cells in the body of an infinite creation which is hurling itself through infinity at unknown speed. Suns, moons, and stars are merely bones in a great skeleton composed of all the substances of the universe. Our own little lives are merely part of that infinite life throbbing and coursing through the arteries and veins of space. But all this is so vast as to be beyond the comprehension of this little "I am" in us. Therefore, we may say that both extremes are equally incomprehensible. We live in a middle world between infinite greatness on one hand and infinite smallness on the other. As we grow, gradually our world grows also, resulting in a corresponding increase in the scope of our understanding of all these wonders.

# THE INFERNAL WORLDS
## Part IV

At the base of the spine is located the throne of the Lord of Form, commonly called Jehovah and Shiva. The *lingam* is His symbol. He rides the great bull of earthiness. His daughter is death and destruction, yet he is not a thing of evil. He builds the bodies which give us the power to function in the lower worlds. He crystallizes them around lines of force. Geometry is the skeleton, and all the bodies which He builds are geometric problems, geometric angles crystallized into rocks and stones. Gradually the crystallization which brings bodies into the world causes them to become too dense and unyielding to respond to the subtle impressions from the spiritual consciousness. Slowly they turn to stone, and death is the result of the same cause that brought the body into the world. The early races of the earth worshipped the procreative attributes of life. They felt that the highest expression of life was the power of giving still another life to the world. Therefore the principle of life giving was personified into a deity who gave life to all things, or rather brought into manifestation the latent life which cannot grow or unfold in the physical world without the vehicle of dense substance.

To the occultist, birth is death and death is an awakening. The mystics of ancient days taught that to be born into the physical world was to enter a tomb, for no other plane of Nature is so unresponsive, so limited as the earth world. Time and distance were prison bars chaining the soul to narrow environments. Heat and cold tormented the soul, age deprived it of its faculties, and all man's life was but a preparation for death. As life is lived under the shadow of death, they taught that it is a mockery, a hollow thing, gilded to the careless eye but tarnished and worm-eaten upon close examination. The physical body became the sepulchre, the tomb, the place of burial, in which spirit lay awaiting the day of liberation when as a virgin spark it should arise again from the broken urn of clay. Therefore in all the religions of the world we have the lower world as a black pit into which the three-headed Yama hurls the souls of the damned to suffer in a hell of their own creation, for it is true that each race makes from its own nature the demons which torment it. Here Typhon, the Egyptian god of destruction, with the body of a hog and the head of a crocodile, awaits with yawning jaws to devour those who have failed to make proper use of life's opportunities. Among most peoples the demon is symbolized as part animal, part human. He dwells in the animal nature of man, and those who are controlled by their appetites, their likes and dislikes, their hates and fears, need no further damnation; they have built their own hell and are experiencing its torments.

The generative system is gradually being absorbed into the brain, and the man of the next great world period will generate its kind, or at least form vehicles for them, through the larynx, which is the organ of the spoken word. We are told that a small etheric body is gradually being built near the larynx, which is later going to form the organ of positive reproduction. Those who are incapable of raising the spinal fire through the Sushumna canal will be cast off into a side kingdom, like the simians of today.

The physical body is supposed to be under the control of the moon, which, as you know, rules all the liquids of the earth. The moon was the last incarnation of the earth spirit, and the human race passed through its state of animal consciousness in the etheric body of the moon lord. The lunar spirits are called "the ancestors" and are known to Christians as angels. These beings have control of the generating powers of animal and man. The life, coming into incarnation often chooses its future vehicle many years before it appears in the world. It is said that the etheric germ is placed in the body of the parent as long as twenty years before the child comes into the world. This is the result of its search for environment especially suited to its spiritual and material needs.

Certain occult schools have taught that the spiritual consciousness of man was not fixed in any point of the body but was in whatever part of man that his thoughts dwelt. We know that there are three worlds where man may dwell. The first is his mental world, where he may live surrounded by his thoughts, his dreams and his aspirations. The second is his human world, where he may be one of the great middle class who think a little, eat a little, sleep a little, and worry unceasingly. The third possible home is his animal world, where he may dwell in the midst of passions, lusts, and hates which burn his soul and consume his body. The history of primitive races shows that they have risen through all of these stages until at last a few have become truly thinking creatures. The blood of every man is individual. When crystallizing, it forms into geometric patterns which differ with every person, so that by means of blood analysis a far surer system could be evolved for crime detection than either Bertillon or finger-prints. The story of man's soul is written in his blood. The position he occupies in evolution, his hopes, and his fears are all imprinted into the etheric forms which flow through his blood stream. Until red blood came into the body, the spirit of man could not enter but hovered over the body attached to it by an electric thread. By studying crickets, grasshoppers, and similar small creatures clairvoyantly, it is possible to observe impulses from the little globes hovering over their bodies which result in the primitive motion and sense which they demonstrate. Therefore, it is said that the actual line between the vegetable and the animal is drawn with the coming of red blood; consequently, certain small fishes, mollusks, etc., are technically vegetables, although not recognized as such by science. The liver is the key to the red blood. Lucifer's red garments derive their color from the blood, while the word *Lucifer* really means "a carrier of light" (or heat) and is a name for the blood. For this reason he is the spirit of temptation. In the Christian Mysteries the piercing of the liver of Christ by the spear of the centurion is especially mystical, while Prometheus, the friend of man, hanging upon the peak of Mount Caucasus with a vulture gnawing at his liver, is the same myth expressed in the symbolism of ancient Greece.

It is further interesting to note the relation between the words *live* and *liver,* for to have a liver is to live. Along the same line we may note that the word *live* spelt backwards becomes *evil,* and the word *lived* becomes *devil.* This peculiar relationship is found not only in English, but to a slightly less noticeable degree in several languages. When we take up this, however, we become involved in the study of Qabbalism, which is the analysis of the meaning of words considered symbolically.

Red is the color of blood and the key of the liver, and its effect upon animals is very noticeable. It irritates, excites and, in some cases, actually causes animals to go mad. Therefore it is often used in making the capes worn during bullfights. These the fighters flaunt in the bull's face, and trouble usually follows. The use of red lights is not uncommon in black magic. Evil magicians use them to materialize spectres, while medical science has already discovered that it is a strong irritant when applied to the human body.

During anger and hate the astral aura of man becomes streaked with red flames very closely resembling thunderbolts. Very often the base of the spine glows with a murky red light, symbolic of hate, passion or anger. This red glow, burning eternally at the base of the spine, has given rise to the story of hell fire and damnation, but the preacher has failed to remind the laity that they carry their own hell around with them wherever they go.

The red power is said to be broken from the white light of the sun through the body of Samael, the spirit of Mars. This is the cause of the red glow in the heavens. Mars is the God of War, wrangling, hate, and dissension. He was the patron deity of the Roman Empire, whose uniformed soldiers wore red as a symbol of his sway. Following the lead of this patron saint, they conquered the world and then fell upon the swords with which they had murdered others.

While red is the color of the body, yellow is regarded as the color of the soul. For this reason the Buddhas and World Saviors are usually symbolized as being surrounded with a golden nimbus or halo. This light is the yellow robe; also the light bearing witness to the darkness, of which St. John wrote. This light, flooding through the third ventricle, represents the Shekinah of the Jews which hovers over the Mercy Seat as a pact between God and man. Yellow is a vitalizer, a life-giver. Therefore the golden-haired sun and its personification—the Christ—are both givers of life. Devitalization may be successfully treated by exposing the spleen to the sunlight.

Blue, the highest of the three primary

colors, is the color given to the Father. It is a very relaxing, restful color, especially valuable in the treatment of insanity and obsession. It is very difficult for black magicians to function successfully in a blue light. Its affinity to the mind is very evident, and it gathers as an electric sea in the pineal gland as an extract from all the spiritual qualities of human nature. The blue heart to every flame was said to symbolize the invisible Father behind the glowing sun. In the words of Christ, "He that hath seen me hath seen the Father. I am in the Father and the Father in me."

The use of colors in symbolism is very interesting. The green dragon, whom the heroes of mythology usually slay, represents the earth. The white armor is a purified physical body. The black magician is darkness and uncertainty. All colors have symbolic value, and great lessons can be learned from the study of the application of these values to occultism.

While discussing the subject of occult anatomy and physiology, we must stop for a moment to give credit to the alchemists and Rosicrucians who, during the Middle Ages, concealed the study of occult anatomy by dressing up the organs of the human body in the form of retorts and alchemical vessels. One of their great exponents said, in substance: "Our chemistry is not with chemicals as you know them, but with certain secret vessels" (internal organs) "and spiritual chemicals which are invisible to the ordinary individual. We do not believe in torturing chemicals," (combining them to form gasses, vapors or seething masses) "for chemicals, like men, can suffer when brought into unkind relationship with each other."

The alchemical furnace was the human body. The fire that burned in it was at the base of the spine. The chimney was the spinal cord up which the vapors passed to be gathered again and distilled in the brain. This was indeed a secret system brought to Europe from the Far East, where for centuries it has been considered the highest form of religion. We may call these occult truths the principles of operative spirituality in contradistinction to modern religion which is entirely made of speculative theories. People do not dream that religion is physiological, nor would they believe that their salvation depends entirely on scientific uses of the life elements and forces within their own bodies; but in spite of all that may be said to the contrary, such is the case. During the next few years much will be done to enlighten man concerning the secret workings of his own parts and members.

It is very interesting to note the similitude

existing between the incarnations or appearances in the world of the great Avatar, Vishnu, and the changes which take place in the human embryo previous to birth. This brings us to our next subject, occult embryology.

# OCCULT EMBRYOLOGY
## Part V

The great Lord Vishnu has already come nine times into the world to save His people. His tenth birth is yet to come. His nine appearances closely parallel the nine principal changes taking place in the human embryo previous to birth. Vishnu was first born out of the mouth of a fish. He then rose out of the body of a turtle. Still later he appeared as a boar, then a lion, afterwards a monkey. And after a number of changes he appeared as a man. I noted some time ago that a scientist had arranged a table showing the relationship of the human brain to various animals during the prenatal period. He followed exactly the list of incarnations of Vishnu, while totally unaware that he was linking together Oriental occultism and Occidental embryology.

The cosmogony myths of nearly every nation are based upon embryology. The formation of the cosmos is said to have taken place in the same way that man is formed only on a larger scale. For example, in *The Vishnu Purana* we are told that creation took place within the womb of Meru. Space was surrounded by great mountains and cliffs (the chorion). The universe was created out of water and floated in a great sea (the amniotic fluid). Down a ladder (the umbilical cord) came the gods. Four rivers flowed into the new land, as told in Genesis. These are the blood vessels of the umbilical cord. So the story goes. A marvelous correlation exists. Some day perhaps a new science can be based upon the law of analogy. This will prove to be a far greater contribution to scientific data than all the scientific speculation of the ages.

It is reasonably certain that the story of Adam and Eve and the Garden of Eden is based upon embryology, and that the womb is the original Garden of Eden. In symbolism it is represented by the "O." The dot in the circle is the primitive germ; and so on as far as you wish to carry the analogy. The Egg of Brahma is the story of the cosmic embryo, and embryology is the basic study of creation.

In embryology we also have a very interesting recapitulation of the passage of the human race through the various species of Nature. Here we find at certain periods the Hyperborean creatures. At another time we see the primitive Lemurian man; later, the Atlantean; and finally, the Aryan. We most certainly recommend to all occult students that they make a very careful study of this subject. Science knows that all life upon this planet came out of the water. The human embryo is enveloped with water through all the primitive stages of its growth, and in it we find the story of the evolution of all things. Sex did not appear upon the planet until the third race. It does not appear in the embryo until the third month.

The recapitulation of the human embryo through the lower kingdoms of Nature is one of the strongest proofs of evolution, inasmuch as it proves conclusively that man could not have been made originally in his adult condition. Consequently he passed through a cosmic embryology; in fact, he is still in the embryo and will not actually be born into the human race until he is truly human, which will not be for many thousands of years. He is just in the state of becoming man.

The nine months of the prenatal epoch have been employed in symbolism for ages. Nine is called the number of man, because of the nine months that the body is in a state of preparation. The perfect number is supposed to be twelve, so at the present time man is born three months before he is finished. The gradual unfoldment of the human race will result in ever more being accomplished during the prenatal epoch, until finally birth will be the ultimate, and all experience and growth will take place in the embryonic state.

Man is not born all at once. We may say that he is born by degrees. The consciousness works outside of the body, laboring with the plastic substances up to the time of the quickening, when it takes hold of the vehicle from within and begins to mold a certain amount of individuality out of the materials which surround it. At the time of birth the physical body is born, and a process of crystallization sets in which never ceases for a moment until death. Man begins to die the moment he is born, and the span of life is the length of time which this requires. At the seventh year the vital body comes into action, and the greatest periods of growth commence. It is then that parents begin to experience difficulties. It is the time of letting clothes both down and out. Children shoot up like weeds, for they are literally recapitulating their plant existences as before that time they were recapitulating their mineral state. At about the seventh year the child begins to manufacture vital essences within its own body. Up to that time it lives upon life forces secreted in the ductless glands of the throat before birth. In other words, it maintains itself upon life which it has stored away from the parent. At about seven it starts work for itself. It is on the go every minute, and if youth could only bottle up its energy and preserve it for old age, what a wonderful world we would live in.

Between twelve and fourteen in the temperate zones the liver starts activity, the emotional body is born. It is during these adolescent years that youth faces its greatest problem. Emotionalism runs riot. The consciousness is recapitulating its animal existences. It may truly be said that these are the days of puppy love. These are years often filled with great mistakes. More lives are blighted between the fourteenth and twenty-first years than at any other period of life. It is especially noted that among primitive races brought in contact with our educational system there is a turning point about the fourteenth year. Up to that time these children are at the head of their classes and extremely brilliant, but when the animal nature takes hold they are absolutely through as far as education is concerned. Any school teacher who has taught foreign children will vouch for this condition among certain nationalities. The moron is an example of the loss of mental function with the birth of the astral body, and there are many of these examples. During these years of emotional riot parents must rule their children with firmness and kindness or those same children will turn some day upon their parents and blame them for ruined lives.

Between eighteen and twenty-one, according to climatic conditions, the mental body takes hold, and we say that the individual has reached the age of majority. He is then permitted to vote; his father presents him with a gold watch and sends him out into the world to seek his fortune. Not one person in a million realizes why twenty-one is set as the age of majority, but every occultist knows the reason. The spiritual consciousness, the true "I am," does not take hold of its new bodies until the twenty-first year. Up to that time it is ruled entirely by the lower sense centers. Life thus progresses in cycles of seven years. As an example of this, we know that the twenty-eighth year is the period of second physical birth; the thirty-fifth year, the period of second vital birth, or, as it is called, second growth; the forty-second year, the period of second emotional birth. During these years people otherwise perfectly normal very often grow sentimental. The forty-ninth year marks the dawn of a new period of mental activity, and the following seven years are the golden years of thought. They are the periods of philosophic reason, the crowning years of life. So on, cycle after cycle. If the individual waits long enough, he may pass through his second, third and fourth childhoods.

Few people realize that they are composed of mineral, plant, and animal elements. The bones are literally minerals, hair is a plant nourished by waves of vital ether pouring out through the skin, while within every individual are thousands of little wiggling, creeping, crawling things that make each of us a zoo all by ourselves. The ancient Scandinavians, realizing this, wrote many legends about the little creatures that lived in man. A famous statue of Father Nile shows him covered over with tiny human figures representing the attributes and functions of man. Man is a great study, but we make very little use of our textbook. The Scriptures of all nations are filled with anatomical references to cities and places that have no existence outside of man himself. The twelve gates of the Holy City are the twelve apertures in the human body. Like the twelve Masters of Wisdom and the twelve great schools of philosophy, these apertures are divided into two divisions of seven and five. There are seven visible openings and five concealed openings in the human body.

One of the Greek philosophers told his disciples to remember distinctly that there were six openings leading into the human brain but only one leading out of the human head, and that one led from the stomach. Therefore he was to listen twice (once with each ear) look twice (once with each eye) sense twice (once with each nostril) but to speak only once, and that what he did say would come from his stomach and not the brain. This advice still holds good.

The Hebrews used the human head as a favorite symbol to express the divine attributes, calling it "the Great Face." The two eyes were correlated to the Father, for they were organs of consciousness; the two nostrils to the Son, because they were organs of sense and also vehicles for Prana, the life force in the ether. The mouth was used to symbolize the Holy Spirit, the one who sent forth the spoken word and formed the world. The seven vowels to which the mouth gave birth were the Seven Spirits before the Throne, also the vials and trumpets of Revelation. They went forth as the army of the voice to create in the seven worlds, and all Nature resulted from their creative power. Few realize the magnificent symbolism concealed within the human head and how it has been used in Scriptural writings.

To this article is appended a treatise which was published separately some years ago but has been out of print over three years. The treatise has a direct bearing upon the subject of anatomical symbolism, showing how the principles outlined in the preceding pages work out when applied to different world problems of today.

## OCCULT MASONRY

To the student of mystic Masonry one problem eternally presents itself. He knows it under many names. It is told to him in many symbols, but briefly it may be defined as the purification and liberation of spirit and body from the bane of crystallization and materiality. In other words, he is seeking to rescue the life buried amidst the ruins of his fallen temple and restore it to its rightful place again as the keynote of his spiritual arch.

When studying ancient Masonry we are dealing with one of the first revelations of what we know as the Wisdom Teachings. Like other great mysteries, it consists of solutions of problems of everyday existence. It may seem of little use to us now to study these ancient abstract symbols, but in time every student will realize that the things he now casts aside as worthless are the jewels which one day he will need. Like the centaur in the zodiac, man is eternally striving to lift his human consciousness from the body of the animal; and in the three-runged ladder of Masonry we find the three great steps which are necessary for this liberation. These three steps are the three grand divisions of human consciousness. We can briefly define them as materiality, intellectuality, and spirituality. They also represent action on the lowest rung, emotion on the central, and mentality on the highest. All human beings are lifting themselves up to God by climbing these three steps that lead to liberation.

When we have united these three manifestations into a harmonious balance, we then have the Flaming Triangle. The ancients declared God, as the dot in the circle, to be unknowable, but that He manifested through His three witnesses—the Father, the Son, and the Holy Spirit. Now the same is true with man. The God in each of us can manifest only through His three witnesses. The Father manifests through our thoughts, the Son through our emotions, and the Holy Spirit through our actions. When we balance our thoughts, our desires and our actions, we have an equilateral triangle. When man's purified life energies radiate through these three witnesses, we then have a halo of flame added to the triangle, in the center of which is God —the unknowable and unthinkable One, the *Yod* or flaming letter of the Hebrew alphabet; the *Abyss* which no one can understand but from which all things come. The life of this Unknown pours outward through the triangle, which in the higher degrees is surrounded by a halo of flame. The halo is the soul built from the transmuted thought, action, and desire—the eternal triangle of God.

Among Masonic symbols is the beehive, called the symbol of industry, for it clearly demonstrates how man should cooperate with his fellow man for the mutual development of all. It, however, contains a much deeper message, for each living soul is a bee that travels through life and gathers the pollen of wisdom from the environments and experiences of life. As the bee gathers the honey from the heart of the flower, so each of us should extract the spiritual nectar from each happening, from each joy, from each sorrow, and incorporate it into the great beehive of experience—the soul-body of man. In the same way it is said that the spiritual energies in man eternally take the life forces he is transmuting and carry them up into the beehive in the brain, where is kept the honey or oil necessary for the sustenance of life.

The ancient gods are said to have lived on nectar and not to have eaten or drunk like other men. It is quite true that honey gained or extracted from coping with the problems of everyday life is the food of the higher man. While we eat at the well-laden board, it would be well for us to consider whether or not the spiritual man is also nourished and developed by the things which we have transmuted in our own lives.

An ancient philosopher once said that the bee extracts honey from the pollen of the flower, while from the same source the spider extracts poison. The problem which then confronts us is: are we bees or spiders? Do we transform the experiences of life into honey or do we change them into poison? Do they lift us (they should) or do we eternally kick against the pricks?

Many people become soured by experience, but the wise one takes the honey and builds it into the beehive of his own spiritual nature.

It is well for us also to consider the Grip of the Lion's Paw, one of the world's most ancient symbols of initiation. In ancient times the neophyte on his way through the mysteries of Egypt's temples was finally buried in a great stone coffer for the dead, later to be raised to life again by the Master Initiate in his robes of blue and gold. When the candidate was thus raised, the Grand Master wore upon his arm and hand like a glove the paw of a lion and it was said of the newly raised disciple that he had been brought to life by the "grip of the lion's paw." The Hebrew letter *Yod* (which is used in the center of the triangle and is sometimes a symbol of spirit because of its flame-like appearance) means, according to the Qabbalist, a hand that is stretched forth. We understand this to symbolize the Sun Spirit in man, which is said to be enthroned in the sign of Leo, the lion of Judah. And as the fruits of the fields and the seedlings are grown and developed through the sun's rays, so it is said that the crystallization of man is broken up and dispelled by the light of the spiritual sun which raises the dead with its power and liberates the latencies of life. The spirit in man, with its eyes that see in the dark, is ever striving to lift the lower side of his own nature to union with itself. When the lower man is thus raised from materiality by the higher ideals which unfold within his own being, it is then said that the spirit of light and truth has raised the candidate for initiation by the "grip of the lion's paw."

It is well to notice the symbolism of the two Johns, as we find them in the Masonic rituals. John means "ram," and the ram is symbolical of the animal passions and propensities of man. In John, the Baptist, dressed in the skins of animals, these passions are untransmuted; while in John, the Evangelist, they have been transmuted until the vehicles and powers which they represent have become the beloved disciples of the Christ life in man.

We often hear the expression, "riding the goat" or "climbing the greased pole." This is of symbolic import to those who have eyes to see, for when man masters his lower animal nature he can say honestly that he is "riding the goat;" and if he cannot ride the goat he cannot enter the temple of initiation. The greased pole which he must climb refers undoubtedly to the spinal column; and it is only when the consciousness of man climbs up this column into the brain that he can take the degrees of Freemasonry.

The subject of the *Lost Word* should be considered as an individual problem. Man himself—that is, the true principle—may be called the *Lost Word;* but it is better to say that it is a certain something radiating from man which constitutes a password recognized by all members of the Craft. When man as the architect of his temple abuses and destroys the life energies within himself, then the builder, after having been murdered by the three lower bodies, carries to the tomb with him where he is laid, the *Word* which is the proof of his position.

Abuse of mental, physical or spiritual powers results in the murder of energy; and when this energy is lost man loses with it the Sacred Word. Our lives—our thoughts, desires, and actions are the living threefold password by which a Master Builder knows his kin; and when the student seeks admittance to the inner room he must present at the temple gates the credentials of a purified body and a balanced mind. No price can buy this Sacred Word, no degree can bestow it. But when within ourselves the dead builder is raised to life once more, he himself speaks the Word, and upon the Philosopher's Stone built within himself is engraved the living name of the Divine.

It is only when this builder is raised that the symbols of mortality can be changed into those of immortality. Our bodies are the urn containing the ashes of Hiram, our lives are the broken pillars, crystallization is the coffin, and disintegration is the open grave. But above all is the sprig of evergreen promising life to those who raise the serpent power and showing that under the debris of the temple lies the body of the builder who is raised *whenever* we so live that the divine life within is given expression.

There are many of these wonderful Masonic symbols handed down to us from the forgotten past, symbols whose meanings long lost have been buried beneath the mantle of materiality. The true Mason—the child of light still cries out for liberation, and the empty throne of Egypt still waits for the King of the Sun who was killed. All the world still waits for Balder the Beautiful, to come to life again; for the crucified Christ to roll away the stone and rise from the tomb of matter bringing His own tomb with Him.

When the man has so lived that he can understand this wonderful problem, then the great Eye or center of consciousness is enabled to see out through the clean glass of a purified body. The mysteries of true Masonry, long concealed from the profane, are then understood; and the new Master, donning his robes of blue and gold, follows in the footsteps of the immortals who are climbing step by step the ladder leading upward to the Seven Stars. Far above, the Ark—the source of life—floats over the waters of oblivion on high; and sends its messages down to the lower man through the cable tow. When this point is reached, the door in the "G" is closed forever, for the dot has returned to the circle: the threefold spirit and the threefold body are linked together in the eternal Seal of Solomon. Then does the cornerstone which the builder rejected become again the head of the corner, and man—the capstone long missing from the Universal Temple—is again in place. The daily occurrences of life are sharpening our senses and developing our faculties. These are the tools of the Craft—the mallet, the chisel, and the rule—and with these self-developed tools we are slowly truing the rough ashlar or cube into a finished block for the Universal Temple. It is only then that we become Initiates of the Flame, for only then does light take the place of darkness. As we wander through the vaulted chambers of our own being, we then learn the meaning of the vaulted chambers of the temple; and as the initiatory ritual unfolds before our eyes we should recognize in it the recapitulation of our own being, the unfoldment of our own consciousness, and the story of our own lives. With this thought in mind we are able to understand not only why the Atlanteans of old worshipped the rising sun but also how the modern Mason symbolizes this sun as Hiram, the high-born, who, when he rises to the top of the temple, places a golden stone upon it and raises to life all things in man.

Printed in the USA
CPSIA information can be obtained
at www.ICGtesting.com
LVHW012234050124
768161LV00018B/1175